The Ultimate Guide to Saving Gas

How To Defeat The "Gas Pump Monster"

With The Vehicle You Already Own!

By

Pete Schultz

Illustrations

by

Dawn Baumer

www.DawnsArtToYou.com

Table of Contents

Legal Notice

Every effort has been made to make this publication as complete and accurate as possible. While all attempts have been made to verify information provided in this publication, the Publisher and or Author assumes no responsibility for errors, omissions, or contrary interpretation of the subject matter herein. Any perceived slights of specific persons, peoples, or organizations are unintentional. In practical advice books, like anything else in life, there are no guarantees of income made. Readers are cautioned to rely on their own judgment about their individual circumstances and act accordingly.

Due to the varying conditions and circumstances, neither Zenco Publishing Co. or the Author offer any warranty or guarantee on any results derived from the information provided in this publication.

This publication contains material protected under the International and Federal Copyright Laws and Treaties. All unauthorized reproduction or use of this material is prohibited.

Introduction

As gas prices continue to rise, with no end in sight, if you're like me, you are looking for anything you can do to deal with an ever-increasing cost of basic transportation.

When asked what to do about it, the first thing that comes to most people's mind is getting better gas mileage. They wouldn't be wrong!

Gas mileage is a major contributing factor in fuel expenses. If one was to dig a little deeper, however, it's not so much a question of getting better gas mileage, it's really more about how to save money on your gas expenses as a whole.

It's much more than just calculating how many miles per gallon you get between fill ups.

This "bigger picture" view will require a slightly different mindset; specifically, one that focuses on reducing your total fuel expenses by influencing the several factors involved in achieving this goal.

The goal of course being, more money in your pocket at the end of the month!

Hi, my name is Pete Schultz. My family has been in the automotive repair business for over 40 years. In this book, I will show you practical ways that you can increase the gas mileage of the vehicle that you already own, as well as disband some of the myths and **"gas mileage voodoo"** out there about how to save money at the gas pump.

Specifically, you are going to understand the concept of gas mileage and gas savings; and how it really affects your finances in the big picture. You'll also learn how little changes can make a big difference in your bottom line.

We will also cover:

- Vehicle maintenance that affects your gas mileage

- Driving habits and techniques that impact your fuel consumption

- Thinking ahead through route planning

- Vehicle modifications

- Tracking your progress

- What to look for when at the pump

- Accessories or devices that "really work"

- A better understanding of fuel-efficient vehicles on the market

At the end of this book you will have some real actionable steps that can make a noticeable impact on your monthly fuel expenses, putting more money in your pocket, and less money in the pump. Not only will you save money in fuel costs, but also in vehicle repairs and maintenance.

Ok, lets get started.

About Gas Savings

As I said earlier, your car's gas mileage is a major factor in the amount of **"cheese"** you have to fork out for travel each month.

If that's the case, Then all you have to do is go out and get yourself a car with great gas mileage and be done with it. Right? Problem solved!

Well, not so fast!

Remember, I said it was **"ONE" factor,** not the "ONLY" factor.

Since our focus is on the Big Picture, the truth of the matter is that a new car may not be the best route to go, and could end up costing more money in the long run.

Okay, you're probably thinking, "Boy, what kind of soap are you selling?"

Let me explain.

Most super efficient cars these days are usually smaller and newer.

Depending on your circumstances, such as the size of your family, or what you have to do with your current vehicle, forking out the cash for a new ultra efficient "skateboard with seats," may end up costing you more money than it would be to keep the one you have.

Here's an example:

Let's say you have a family of five. In this case having a vehicle that's big enough for everybody to travel in would be cheaper than having to make two trips or taking two smaller vehicles. No matter how good they were on gas mileage.

Having a good idea of how you will be using your vehicle is a major factor to consider. Have you ever tried to haul a load of wood in a Ford Focus? Believe me, it doesn't work out very well.

Don't get me wrong, there are situations where having a small efficient vehicle would be beneficial. For instance, if you don't have a large family or you have to make long commutes to work. In this case a small efficient vehicle would be perfect as a primary or a second vehicle.

Another thing to consider is that the cost of insurance and repairs of a second vehicle may outweigh the gas saving benefits of getting one.

The most important thing in my mind is safety!

It's a Scientific Fact!

It is safer to drive a larger car than it is to drive a smaller one.

There is no substitute for "**Real Metal and Crush Space**" when involved in a collision.

One of the first things car manufacturers must do in order have a vehicle get better gas mileage is drop weight. They end up replacing good old-fashioned steel, with plastic and other lightweight materials.

Sometimes, having to pay an extra $50 a month in gas may be more than worth it. Especially when it comes to the safety of your loved ones.

The Dawn Of A New Age

Being in the automotive repair industry for many years, I have witnessed many outstanding transformations and technological breakthroughs in the industry.

The most significant breakthrough in automotive engineering in the last 20 years has been the introduction of **Electronic Fuel Injection** (EFI) **and Electronic Ignition Systems** (EIS)**.**

EFI and EIS systems have turned the automobile into a "**dynamic mechanism",** that can monitor and adapt to changing conditions through the use of a Power-train Control Module or "PCM".

These "computer controlled" systems are directly responsible for doubling or even tripling the expected lifetime of a vehicle's engine from 30 years ago.

Despite what you may have heard, or led to believe, almost any vehicle manufactured after 1989 that's running properly, can achieve an almost perfect combustion.

As a result, we have cars that have been on the road for years with drastically fewer gas emissions and greatly increased fuel economy than the cars built in the 80's.

This breakthrough didn't come easy, believe me, I know!

I had the challenge of trying to fix those cars that they were using as guinea pigs in the early 80's.   But they got it! Thank God.

So, is there some kind of a moral to this story?

Yes, yes there is!

Rushing out to buy a new electric vehicle that requires a "Coal-Fired" Power Plant to keep it charged, all in the name of saving the planet, may not be the most effective way to do it.

Meanwhile, through recent technology and changing your own habits, you can make a real impact on both the planet and your wallet.

All with the vehicle you own, Right Now!

Understanding Gas Mileage And The Effects On Your Wallet

To get started, we first need to have a good understanding of exactly how gas mileage and fuel costs affect your wallet, and how little things can add up to big savings over time.

First off, miles per gallon (MPG) can be very misleading as a gauge for how much fuel you actually use.

For instance, MPG is completely dependent on the conditions that the vehicle is experiencing at any given point in time.

If you drive up a hill for 100 miles you may only get 10 miles per gallon. When driving down a hill for 100 miles you may get 60 miles per gallon.

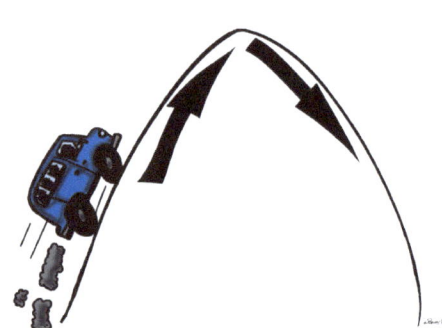

The whole miles per gallon rating is completely irrelevant! This makes actually interpreting the effects of miles per gallon very difficult.

In reality, just small increases in fuel efficiency, even in cars that are not very fuel-efficient, can make a big impact on your total fuel costs.

How about an example:

Let's say you have a car that gets 12 miles per gallon. And also let's say you are able to increase that to 15 miles per gallon.

If you drove 10,000 miles in one year, that would be a savings of approximately 170 gallons of gas. At $4.00 per gallon, that translates into about $700 of cold hard cash that you saved on your yearly gas expense.

As you can see, just a mere 3 miles per gallon increase with the vehicle you already own, can add up big.

Now that's just on miles per gallon alone!

What if, by implementing some changes in your daily habits and some simple route planning, you could shave off an additional 3,000 miles per year of use?

This could add an additional $200 dollars in savings on gas alone, not to mention, reduced vehicle maintenance.

How much does a brake job cost?

One of the first habit changing techniques you will learn is how to change your way of braking. This alone can save real money when it comes to the repair shop.

Remember our focus is on "**BIG PICTURE**" money savings.

It truly is a bunch of small things working together that add up to big savings at the end of the year.   We're talking "real money" that can go towards food, paying off credit cards, or even a nice vacation.

Things That Affect Gas Mileage

Temperature

Cold temperature in particular effects gas mileage. Cold air is denser than warm air and it takes more gas to make up this difference. Sub zero temperatures greatly reduce your gas millage. Unfortunately there's not much we can do about that one except move to a warmer climate.

Auxiliary Lights

Many manufacturers have their car's headlights stay on all the time. If you have the ability to turn them off, I would. Only use your headlights or "driving lights" in adverse conditions such as rain and snow, dark tunnels, or when lighting requires you to do so.

Defrosters

In cold climates, it is customary to start your car ahead of time to defrost the windows. Although it's important to let your car's engine warm up a little before you drive off, letting it run for an extra 10 minutes in the morning just to get it warmed up uses lots of gas.

In addition, electric defrosters on the rear windows take up a lot of energy compared to the old-fashioned ice scraper. The same goes for accessories such as seat warmers.

Believe it or not! The simple act of parking your vehicle inside a garage can save you money. Vehicles parked in a garage stay warmer, don't need defrosting and take less time to warm up.

Are you starting to see a theme catching on here?

Basically it is physics, and it works like this:

Any amount of extra energy your car requires, whether it be from heating elements, extra lights, or fan blowers, translates into extra fuel. Even a weak car battery can cost you money and fuel.

Here's an example:

If you've ever used an outdoor generator, you may have noticed that the more things you plug into the generator the harder the motor seems to work.

The generator on your car is the same way. The more power it has to provide, the harder your car's engine has to work.

Like I mentioned earlier, your car battery can cause an extra load on your cars engine. This is do to the fact that batteries lose their efficiency over time. When you start your car, it takes a lot of power from your battery. Your car's generator immediately goes to work to charge the battery backup. Old, worn-out batteries take considerably longer to charge than newer ones.

If your battery is several years old, you should consider replacing it.

Air Conditioning

There has been a lot of misinformation about using the air conditioner.

Should you or shouldn't you use it?

The fact of the matter is:

It takes a lot of energy to power your car's air conditioning pump, and running it can reduce your vehicle's gas mileage by 10 to 20%.

As a general rule:

Don't use the air unless you have to.

I know there are some areas in the country that this is not an option. There are however, some things you can do to reduce the need to run it as often.

Things like:

- Parking your car in the shade or in a garage to reduce heat buildup.

- If you're planning on going somewhere, open all the windows a half hour before you plan on leaving, allowing the car to cool down.

- Drive with the windows down.

Driving with your windows down does increase wind drag, which can reduce your gas mileage. If possible, use your car's vents instead. Either way, driving with the windows down uses less gas than running the air conditioner.

The only exception, is when traveling at highway speeds.

At highway speeds, not only is it just plain annoying to have the windows down, but the negative effects of wind drag starts to take over. In which case, it may be better to go ahead and use the air conditioner.

Once again, using the fresh air vents would be best.

Another really good tip is to plan your travels in the morning when it's cooler out. This again reduces the need to run the air conditioning.

Cruise Control

Cruise control works really well on long flat stretches such as highway driving, but is very inefficient when it comes to hills or curves.

Later in this book we discuss techniques while driving in "hilly country". At time's like this, you will be much better off being in control of the accelerator pedal.

Turning Off Your Vehicle

There's been a lot of talk about turning off your vehicle. They say "if you must sit idle for any more than 1 minute, turn off your car." **Personally, I think this is Ridiculous!**

Granted, if you find yourself in a traffic jam, or waiting for a train, turning off your vehicle makes some sense. But constantly turning off and re-starting your engine "**WILL**" put excessive wear and tear on your car's starter, flywheel, and battery, not to mention your ignition switch.

Since this is a "practical guide" for saving gas and money, I recommend you use a little common sense when applying this technique.

Vehicle Maintenance

As mentioned before, virtually all vehicles manufactured from 1989 and later, can achieve near-perfect combustion and good gas mileage if running properly.

That's the key - "If running properly!"

If your goal is to save money on your gas expense, you first need to make sure your vehicle is running as efficiently as possible.

Regular vehicle maintenance is a must!

Not only will it save money in gas, but will also extend the life of your vehicle.

Some maintenance things are easy to do, and should be done regularly.

Others are more complicated and should be performed by a certified mechanic.

- First and foremost, you need to **check your tire pressure**. Make sure they are running at factory specifications. Tires that are low on air take a lot more energy to roll and can be dangerous at higher speeds. Over inflated tires can also be dangerous and will wear out the tires prematurely.

- **Change your oil and filter often,** at least every 4,000 miles.

- **Use synthetic motor oil**. Synthetic motor oil protects your engine longer and reduces friction better than conventional oil can. Although synthetic motor oil is a few dollars more per quart, the added benefits well outweigh the cost.

- **Change your air filter often**. Examine it at least every three months. A dirty air filter is a prime cause of poor gas mileage and is easy to replace. There are performance air filters that have been proven to increase gas mileage which we will discuss later in this book.

- **Regular tune ups are a must!** Spark plugs, ignition wires, PCV valves, and fuel filters all wear out over time. As they wear out, your vehicle begins to run less efficiently.

- **Have your brakes and axles checked often.** Your brake pads actually skim right along the surface of your car's brake rotors which are connected to your wheels. A bad brake hose or even just dirt and grime can make them stick and actually drag on your car's brake rotors. Not only will this wear out your brakes very quickly, you can imagine what it will do to your gas mileage as well.

- **The same goes for your universal joints and CV axles.** Many times these joints do not look bad but are actually wearing out inside. This wear does not allow them to move freely and takes more power from your engine to propel them.

"If you feel a vibration either when pushing the accelerator or when coasting, these are often the cause."

"Any time you see your check engine light come on, or you notice a different smell or odor when you start your car, it is a telltale sign of a problem and should be checked out immediately. "

"Excessive smoke or a rotten egg smell coming from your exhaust pipe are signs that there is a problem with either a leaking fuel injector or a bad oxygen sensor. Any one of these components malfunctioning can devastate your gas mileage, not to mention the performance of your car."

On top of good old-fashioned tune-up parts, there are many new products available to your automotive repair shop, such as additives placed in the fuel system and the engine block of your vehicle.

These products are very powerful and should only be used by a certified mechanic. They do a very good job of cleaning out your engine and fuel system from all the deposits that build up over time.

I would highly recommend having this done.

Driving Habits

Now that we have your vehicle running in tip top condition, it's time to get down to the **single biggest thing** you can do to save money on gas and increase your MPG.

Let's take a look at the way you drive!

You may not realize it, but nearly everything we do while driving has a direct effect on our gas consumption. **"We are all creatures of habit, and the way we drive is no different."**

Making a few subtle changes in our daily driving habits, and just being "conscious of how we drive", is where you can make the biggest difference in your fuel expenses.

You need to be aware that there are "laws of physics" at play as you drive your car.

You'll probably remember this from high school science class (that is of course, if you were not sleeping at the time). The law goes something like this:

That which is in motion, tends to stay in motion.

Simply put, it takes less energy to keep something moving that's already moving.

If you've ever ran out of gas, and you ended up pushing your car, you may have remembered how hard it was to first get the car to start moving. But once it got rolling, it was a lot easier to keep it rolling.

That is the key to getting better gas mileage.   "Keep the Momentum."

Every time we have to use the brake to slow down, momentum is lost. That loss of momentum has to be made back up with the gas pedal.

Being aware of your surroundings, and simply thinking ahead of upcoming traffic conditions, is a key factor to keeping your car's momentum, and avoiding unnecessary braking.

I'll say it again!

"Every time you use the brake, you'll have to use the gas."

A good example of planning ahead and being aware of your surroundings is when driving on the highway.

Say you're going the speed limit on the highway and you notice that you are coming up on a slow moving truck. Instead of waiting until you're upon the truck to see if you can change lanes, (which often results in having to slow down and wait for the lane to clear) you should begin to look for openings in traffic ahead of time, so you can just merge into the next lane without having to lose any momentum.

This brings me to another concept that we'll discuss in detail later; **that is speed.**

If you're traveling faster than everybody else you will have to swerve and change lanes more often. You're better off in the long run to match the speed of the other vehicles, and just keep it nice and steady.

Another good practical application of this concept is if you're traveling down a street and you see that there is a red light ahead. Start slowing down by lifting your foot off the gas and begin coasting. The idea here is try to give the light a chance to change before you get there. This often prevents you from having to come to a complete stop and lose all of your car's built up momentum.

Accelerating

We already realize that anytime we step on the accelerator, it costs us money. So it only makes sense that if we change the way we use the accelerator, we can save money. This is an important concept that once again, boils down to "**personal habits**."

Once again we refer to the Laws of Physics:

"It takes the most energy to get the ball rolling."

In other words, it takes more gas to get up to speed than it does to keep it.

"Lets face it, many of us just have a habit of lead-footing it."

Let's pretend for a minute:

You're at a stop sign. As you take off, your goal is to get your vehicle back up to the speed limit, in this case 45 mph.

There are two ways you can get your car up to the speed limit.

- You can quickly press the gas pedal about ½ way and get up to speed quickly.

or

- You can gently press the gas about ¼ way down and glide up to speed.

"In either case, you reach your desired speed of 45 mph."

Now if you gun it, you will get up to speed in less time. You'll also use twice as much gas to do so.

If you take the gradual approach, it may take you an extra 30 seconds to get up to speed, but you will do it with only a fraction of the fuel.

Braking

Now that we are aware of how our accelerating habits can affect our gas mileage, let's dive into the other side of the equation; "Braking".

With a little focus and practice, you can make changes in your braking habits that will make a **"Big Difference."**

"It is as simple as just being more aware of when it's time to brake."

Most of us have our foot on the gas until we have to put our foot on the brake!

Remember!

Any time we have our foot on the gas, it costs us money! Anytime you put your foot on the brake, you're losing momentum! That lost momentum has to be made back up! Which, you guessed it, costs money!

What about when you're not pushing on the gas or the brake?

When your vehicle is coasting, it's using its built up momentum to carry it along while the engine is idling. This uses very little gas. The goal is simple, **"coast more often."**

Here's the best way to achieve this; not only will it save gas, but it will extend the life of your braking system.

If you know there's a stop sign coming up, remove your foot from the accelerator sooner than you normally do, allowing you to coast to a stop.

This **"Will"** take a little bit of practice at first.

If you find yourself still having to brake rather hard when you reach the stop sign, this means you need to let your foot off the gas even sooner. And of course, if you have to get out and push a car to the stop sign, you have to start coasting a little later.

This one technique is HUGE. Think about it!

How many times a day, week, month, do you start and stop your car at street signs?

Say you let off the gas 100 yards sooner than you normally do when stopping.

How many hundreds of yards will that add up to at the end of one week?

That's thousands of yards that you would've been pushing on the gas instead of coasting. This really adds up over the course of a year.

Stop Signs

Here's another tip that works well with stop signs. Once again, it comes back to being conscious about what's going on in front of you.

If you see there are cars at a stop sign ahead of you, just like the street light scenario mentioned before, begin coasting ahead of time. What you want to do is allow yourself to slow down naturally. By beginning to slow down ahead of time, you will hopefully give the cars in front of you time to pass through the stop sign. So ideally, when you get there, you will only have to make one stop.

I know I sound like a broken record here, but remember the **Motion Theory**? How it was harder to get a vehicle moving than to keep it moving?

The same deal applies here!

If you have to stop three car lengths back form the stop sign, then step on the gas to move up one car length, then start again to move up an additional car length, you're just repeating the hardest thing your car has to do, **"Starting the Momentum."**

So by allowing time for vehicles to get out of your way, you will eliminate all that extra stopping and starting.

Road Curves

You want to handle curves in the same fashion. Pay attention to road signs, and if the sign says a curve is coming up, let your foot off the gas and coast as you slow down to handle the curve, verses running up on it and having to apply the brakes.

Peaks and Valleys - Hill Driving

"When in hilly territory, you want to use the roller coaster effect."

Roller coasters don't have any type of motor in them. They rely on gravity and momentum. You can use the same principles when driving on hilly roads.

If you see a hill coming up in the distance, try to gradually increase your speed while still on flat land before you reach it.   This increased momentum will help carry you up the hill without having to use as much energy or gas as you would if you were to wait and have to floor it.

If you have multiple hills, you will want to use gravity to your advantage. To do this, when you are going down one hill, allow gravity to help your vehicle accelerate.

By doing so, you'll be able to increase your speed quite a bit without using very much gas. That extra speed will help you climb the next hill with less effort. Of course, you will want to pay attention to posted speed limits as you use this technique.

Wind and Speed

There is a direct correlation between how fast you go, and how much wind resistance your car has to overcome.

There are two types of wind resistance.

- First, there is the wind that your vehicle has to push out of the way as it propels itself forward.

- Second, there is wind resistance in the form of drag. Drag tries to suck the vehicle from going forward.

An example:

If you're pulling a big tall trailer, one that sticks 6 feet up above your truck, that 6 feet is pushing against the air like a snowplow pushing against snow. The faster you go, the harder your car has to push.

When you have your windows down at higher speeds, the open windows actually inhibit the air from sliding by your car, and tries to suck it in as it goes by.

This is called **"Wind Drag"** and is similar to dragging an anchor off the back of a boat.

Each vehicle has a certain "design characteristic", where the speed and air resistance relationship come into balance.

What this means is that there are certain speeds that a car can go where the air resistance does not effect it very much. On most cars this is **less than 55 mph.**

If you start going faster than this "balance zone", the amount of power it takes to go over that point expands dramatically.

Think of it as if you have one of those big exercise rubber bands.

Step on one end and hold the other at about waist high. It is not very hard to do. But if you raise it up over your head, there is more resistance and it becomes much harder to hold it there.

Studies show that just a 5 mph increase can use up to 1/3 more horsepower to overcome increased air resistance. So you will use less gas going 65 mph vs 70 mph.

Now back to Drag. As I mentioned, open windows create a lot of drag and should be avoided at higher speeds if possible.

There are also other things that create drag on your vehicle. Anytime you attach something to the luggage rack you create a lot more drag, so this should be avoided.

Always try to stow your gear inside the vehicle instead of on top of it.

The luggage racks themselves can create a bit of drag. Most of them have thumb screws that allow you to quickly remove them. If you don't plan on using it, remove it until you actually need in it.

Open sunroofs create extra drag also. There are air deflectors on the market that can be installed to help reduce this effect.

Another factor of speed has to do with "**Timing**".

Street lights normally operate on a timing system. They are timed to keep traffic flowing at the posted speed limit.

If you go faster than the speed limit, more times than not, you will catch more red lights. The same is true if you go too slowly.

So the next time you're at a stoplight, when it turns green, try to maintain the posted speed limit as consistently as possible. By doing so, you will most likely avoid having to stop at every single stop light, and eliminate countless stop and start scenarios.

Lose The Weight

Have you ever noticed in those pirate movies that whenever they have to outrun somebody they start throwing off their cargo and heavy cannons?

Or in action movies when the plane is running out of gas and in order to make it to the landing strip, they always start throwing out the seats and the luggage. (Or the people they don't like, depending on the type of movie you're watching!)

Well your car is the same way!

The heavier your car is, the more power it takes to push it, and the more gas it uses.

"Some people take this to the extreme!"

They pull out the door panels, floor mats and unbolt the front passenger seat. Yes, all this will help your gas mileage. But let's face it, that's a little **Ridiculous.**

However, there are a few practical things you can do to make your vehicle lighter.

Without Going Overboard!

Let's take a look at your trunk!

Basically, you want to remove anything that doesn't need to be there. That may include heavy speaker systems, your rock collection, or maybe a toolbox. Anything you can live without, because it all adds up.

Although I don't generally recommend it, your spare tire does weigh quite a bit, and can be easily removed. This may be a good option if you only drive the car locally, and you always have your cell phone on you to call for help if needed.

"If you're going out of town, I would leave the spare tire in place."

Another good area to shed some pounds is the extra seats. Many Minivans come with several seats that are designed to be easily removed. If you don't need them, get them out of there. You can always throw them back in when you need them. In the meantime there is no sense in hauling around the extra weight.

What "Drive Gear" Should I Be Using?

I have seen people over the years at our shop talking about how they never use overdrive! They think they are saving gas or extending the life of their drivetrain by not using overdrive. **This is false!**

Automatic transmissions found in cars made after 1990 are designed to be operated in OVERDRIVE, except when towing. The car's power-train control module monitors speed, throttle position, air flow, and a whole host of other things and will best determine whether on not the car shifts into overdrive gear.

So what is Overdrive?

When I first heard of overdrive, I thought it was some type of "**rocket engine**" or something cool like that.   But in reality, is just a term for a gear ratio that has a higher output than its input.

In most cases, the drive gear or the D on the steering column, is a 1-to-1 ratio with the motor. Meaning, if the motor has an output of "1", the transmission has an output of "1".

With Overdrive, that output of "1" from the motor is being multiplied by the transmission to say an output of "1.5".   It's going "OVER" drive.

A real easy way to demonstrate this is to use a 10 speed bike.

In 1st gear, it is super easy to peddle. You can peddle fast, but you don't find yourself going anywhere quickly. Put it in 5th gear, you're peddling slower, but you're going faster. Now if you continue pedaling at the same speed, and put your bike in 10th gear, you will be going a lot faster.

Your car's "Overdrive Transmission" works the same way.

If you drive a stick shift, you want to drive in the highest gear you can, without bogging down the motor at slower speeds.

Route Planning

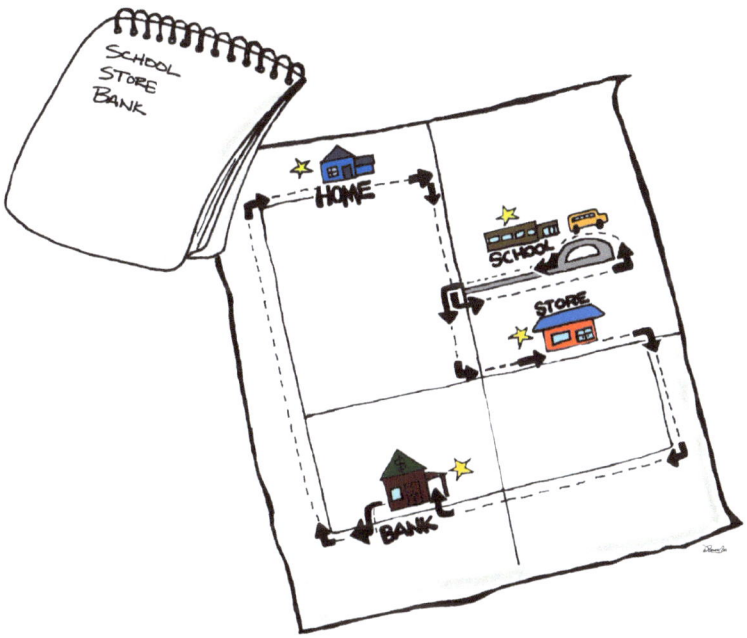

Up till now, we've covered:

- Simple things you can do to your car.

- Changes you can make in your driving habits.

- Various things that have a direct effect on your fuel consumption.

Now let's move into the "Third Pillar" of saving on your gas expenses.

How Often, and Where you Drive!

With a little bit of planning, you can make that tank of gas last longer.

Consolidate Your Travels

This is a big one, but it's easy to do. Just take out a calendar or a notebook and jot down all the errands that you'll have to do for the week or even just the day.

Now, look at your schedule and see what tasks you can combine into one trip. Preferably one that you have to go on anyways, such as taking the kids to school or a doctor's appointment.

When gas was a lot cheaper, many of us would just jump in our car and go get what we needed, when we needed it, and not think twice about it. But these days, $20 in gas doesn't go nearly as far. You would be surprised by how much a little planning can make a big impact on your pocketbook.

Next, let's look at the routes used on a normal basis, like back and forth to work.

When planning your route, ask yourself these questions:

- Am I taking the shortest route to my destination?

- Does this route involve a lot of stop signs or streetlights?

- Are there a lot of hills or curves that need to be dealt with?

GPS Systems

Most people have a GPS system. Many cars come with them and many cell phones have them too. They are great, and can really help you to save gas.

Not only can they tell you the shortest route, they also help keep you from driving around in circles, wasting gas, trying to find your destination.

Make sure that you check the settings on your GPS to display the shortest route, not necessarily the fastest route. Many times the fastest route is several miles longer than the shortest.

Note of Caution! At times, the shortest route is not exactly the most pleasant one to drive. It may be full of turns and hills as well as stop signs. Be sure to analyze your GPS results to determine the best route. If you don't have a GPS you will find a list of some good ones in the resource section of this book.

Avoid Turning Left

Try to plan your route to avoid making left-hand turns.

Did you know that UPS plans their routes so they don't have to turn left?

Usually every time you have to turn left, you have to wait for oncoming traffic to clear. Further more anytime you are standing still with your car running, you are getting **0 miles per gallon**. To make matters worse, you often have to "**gun it**" to beat oncoming traffic.

On the other hand, when turning right, you can usually do it on a red light, and is fairly easy to merge into the flow of traffic.

Straighten Up!

Remember in geometry class?

The quickest way between two points is a straight line, so always look for the most direct or straightest route when planning your trips.

Call Ahead

This one may seem silly but you'd be amazed how often this happens.

Don't you hate it when you go to a store only to find out they're sold out of what you're looking for? Or go to a friend's house to find out they're not home? Little things like this can add many unnecessary miles to your travels.

Try to avoid unnecessary trips by calling ahead.

When Shopping

When you're at the store, try to avoid driving around the parking lot three or four times looking for the perfect place to park.   Just grab the best spot you can quickly, and walk the rest of the way. It will probably do you some good.

The same goes for fast food restaurants. How many times have you waited 10 minutes in the drive through line, and when you finally got up to the window, you noticed that hardly anybody is inside? If there is any kind of line in the drive-through, instead of idling in line, park the car and go inside.

Expense Tracking

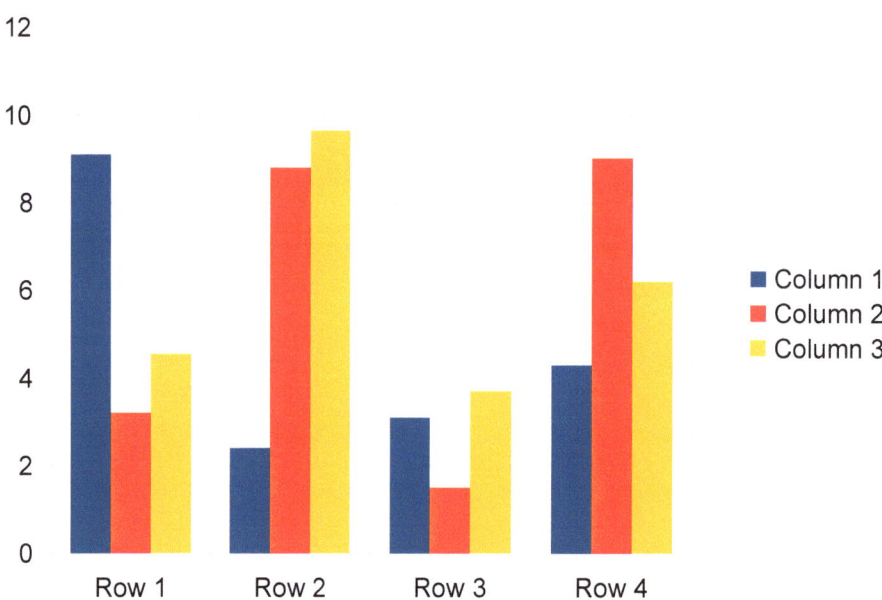

By now, you have a ton of knowledge and know all kinds of cool tricks to get better gas mileage and save gas. So how do you keep track of your progress?

Establish a baseline figure to compare your progress to.

You need to know what your monthly gas bills are currently, as well as what kind of gas mileage your vehicle is getting now.

As for your monthly gas bills, a notebook works very good for this. Round up all your receipts for gas for the month and just add everything up. You could then divide this figure by 30 days if you wanted a daily average.

Your monthly expenses will fluctuate as the price of gas does, but this will give you a good baseline average of your monthly fuel expenses.

Just keep in mind that if gas is 10% higher that month on average so will your fuel costs. In that case you can just subtract 10% from the total to give you your baseline figure for easier comparisons.

Figuring Your Miles Per Gallon - MPG

In order to figure out your vehicle's gas mileage, the best way is to run your tank as low as you can, then fill it up completely.   Next, reset your tripometer to zero.

If you're not sure what the tripometer is, it is a separate setting on your odometer gauge. The odometer keeps track of the total miles your vehicle has driven. The Tripometer keeps track of the amount of miles you have gone in a particular time frame.

At this point, you don't have to worry about how many gallons you put in the tank. Just as long as it's filled up as far as you can get. Now, just drive like you normally do until the tank gets down to about as empty as before.

When you fill the tank back up, write down the amount of miles that the tripometer shows you went, then reset it. At this time you also want to write down how many gallons of gas it took to fill your car back up.

Now that you have these figures, grab a calculator and do this simple equation.

Take the amount of miles you've driven, let's say 200 miles.

Now divide that by the amount of gallons it took to do it.

For this example let's say 17 gallons.

200 / 17 = 11.76 miles per gallon.

So you got 11.76 miles for every 1 gallon of gas used.

Take these figures and write them in a column in your notebook.

By keeping these figures in your notebook you will be able to spot trends in how much gas you use and notice if things change unexpectedly, which may indicate a possible malfunction with your car.

The next time you put gas in your vehicle, just write down the amount of miles driven, and divide that by the amount of gallons you put back in the tank.

Be sure to reset your tripometer each time!

This is the Easiest way to figure your vehicle's gas mileage.

At The Pump

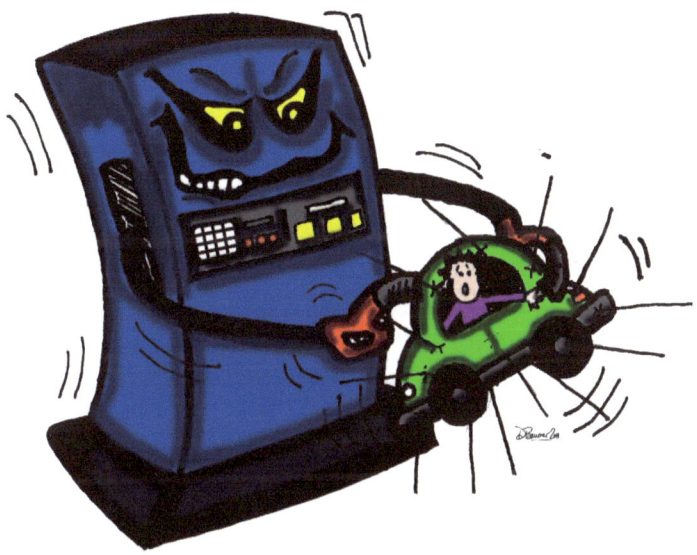

Just as things can be done to your car, and the changing of your habits will affect your gas mileage, there are several things you can do at the pump.

Grade Of Gas

Except for a few high-performance vehicles, all cars manufactured in the US are engineered to run on low octane fuel. In fact, all fuel at the gas pumps have to meet minimum EPA fuel standards. So there is no need to pay the extra $.30 or more for the premium gas unless you experience a noticeable performance reduction.

If you do, simply move up to the next octane number. Nine times out of ten, your car will run just fine. In fact, most cars on the market today would not even benefit one bit from premium gas, so save your money.

When To Fill Up

Try to buy your gas in the morning. That way you can get a nice cup of coffee at the same time. Ok that's not the only reason, the real reason is that the temperature is much cooler. Colder temperatures make fuel denser, meaning that you will get more fuel than the pump is charging you for.

When deciding when to buy gas, you should let your tank get at least half empty. You want to avoid just topping off the tank because many gas pumps have to run a short amunt of time before they really activate and start to dispense gas accurately. So if you just put in a few gallons here and there, there's a chance that they are short changing you. In addition, it's harder to determine your gas mileage on such small amounts.

Where To Fill Up

Always try to buy your gas from a busy station; one that has a lot of customers. This means that their underground tanks are being refilled often and not allowed to go stale.

On the same token, don't fill up your tank when you see the fuel tanker truck filling their tanks. When a gas station's tanks are being filled, all the impurities that are usually on the bottom get stirred up. You don't want that inside your tank.

Whenever possible, try to avoid using Gasohol.

Gasohol is gasoline with alcohol added to it. It's a great idea; except for one problem. It doesn't produce the same amount of power that pure gasoline does. Burning gasohol will reduce your gas mileage.

Gasoline Credit Cards

Many gasoline companies have their own credit cards, and often offer discounts on gas just for using them. This is a great way to get additional savings on your gas as long as you pay it off each month and don't incur unnecessary interest.

Tires

The type of tires your car has can effect gas mileage.

Meaty snow and off-road tires are designed for "TRACTION", not gas mileage.

Today's tire manufacturers, have a wide variety of tires that are fuel-efficient. If possible, try to keep the same size tire that came from the factory.

Tire size is a major factor when a vehicle's computer system is calibrated at the assembly plant for maximum performance and fuel economy.

I know I mentioned it once before but it is very important.

Make sure your tires are inflated to the proper level at all times.

You should check the pressure of your tires at least twice a month.

Temperature makes a big difference when it comes to tire pressure! The colder it gets, the air in the tires contracts and the flatter they become.

Visit the resource page for more information on tires.

www.HowToGasMileage.com/resources

Accessories And Devices

We've covered a lot of information! You've learned some rock solid things you can do to cut your gas expenses. Now it's time to gear up and dive into the seemingly endless pit of products and devices that claim "**unimaginable**" increases in gas mileage.

From **"Harnessing the Moons Power,"** to the magic **"Gas Pump in a Can!"**

No matter what the claims are on the TV and in magazines, consumer studies have shown that most of the products on the market either don't work as claimed, or have no significant benefits compared to the cost.

There are however, products on the market that have been proven to work and have worked for years. I will now introduce some of these products and help you separate the **"Voodoo"** from the **"Should Do"**.

Additives and Potions

I'm sure you've seen the ads "It's the miracle gas pump in a can! Just one drop will get you will get 1,000 miles to the gallon!"

For the most part, none of the additives are worth the money. The only additive proven to have any benefit would be octane boost. This is not really a gas mileage thing, but more of a performance thing.

Acetone

There is a lot of talk about adding acetone to your gas tank. Some people swear by it! Some people say it doesn't do a thing!

Don't add acetone to your gas!

Whether it works or not, your fuel system contains various rubber and plastic parts, that can be damaged prematurely by adding chemicals to your gas. This damage may not show up for years! It can mean hundreds if not thousands of dollars in repairs.

Gasoline manufacturers already put additives in their fuel as needed.

Oil Additives

There are some oil additives that **"DO"** make a difference in your gas mileage! Not only that, they can extend the life of your engine as well. They do this by reducing friction and heat. Most quality synthetic motor oils have these additives already in them.

Visit the resource page for a list of oil additives.

www.HowToGasMileage.com/resources

Gadgets

There are many gadgets out there these days. One of the most popular are magnets.

Basically, they are attached to the fuel line, and are said to change the molecular structure of the gas. Most cost between $15-$40 and in my opinion are not worth the time.

They don't do any harm, and are easy to install, but on average, you may save a 10th of a gallon between fill ups, if anything at all.

Now let's get down to stuff that can really make a difference!

Motor Oil

As I mentioned before, Quality Synthetic Motor Oils are good.

Performance Air Filters

Performance air filters can make a difference in both horsepower, and fuel savings. Your engine needs air to burn the gas. The easier the engine can get the air it needs, the more efficiently it can burn the gas.

Air cleaners can be responsible for up to 10% of power performance and fuel consumption. This is why it is so important to keep your air filter clean at all times. Performance air systems come with washable air filters so you only have to buy them once and are easy to install.

Visit the resource page for a list of Air Cleaners.

www.HowToGasMileage.com/resources

Performance Exhaust Systems

If making it easier for an engine to get air increases performance, how about getting rid of it? Would that help too?   **Absolutely!**

This is what we call **"Letting your Engine Breathe"**.

Here's a perfect example.

Try running a marathon while breathing through a straw. Unless you are some kind of freak of nature, you will not do very well, because your breathing is being restricted by the straw.

Performance exhaust systems are specialty designed to let your car breathe. Several manufacturers offer complete performance exhaust kits that are easy to install.

When combining the benefits of both a performance air cleaner and exhaust system, you start multiplying the benefits, resulting in better engine performance and overall gas mileage.

Visit the resource page for a list of Performance Exhaust Systems

www.HowToGasMileage.com/resources

Chips and Programming

This is where it gets really exciting!

Remember in the beginning, how I explained that the invention of fuel injection, electric ignition, and computer control, has changed the automotive industry?

Here's Why: By having all major components controlled by a computer, it allows for immediate adjustments to the current conditions, providing maximum efficiency. The car's computer does this through the use of it's programming.

Before this, you had to physically change your engine's tuning.

Example:

If you are driving in a cold or a high altitude climate, your car would have to physically be tuned for those conditions. Meaning that you would have to break out your toolbox, take apart the carburetor, figure out what size jets you need to get just the right amount of fuel and air, and re-install them.

Then, a week later you decided to take a trip down the mountain, and are now in a desert climate. Your car would not run efficiently in this new environment because it was tuned for a different one.

With **"Computer Control"**, all adjustments are done automatically,

Your car's computer or PCM (power-train control module) deals with these variables through its programming. The programing is stored in a removable chip called a PROM.

Just like your home computer, it has hardware, and that hardware is controlled by a program. Different programs can operate the same hardware in different ways.

In general, car manufacturers have to design their cars to meet the all around needs and conditions of the mass population. As well as comply with government regulations.

The fact is, it's virtually impossible to have one vehicle do everything.

There is a reason that a dragster that can reach speeds over 200 MPH gets almost ZERO miles to the gallon. And a moped that can only go 25 MPH, can get 150 miles to the gallon.

Now the really cool thing is that you can purchase "**Performance Chips**" that changes the computers programing to one that better fits your needs.

Believe it or not dealerships do this all the time!

Just like your computer downloads updates for its programming, dealerships often reprogram a car's computer to fix problems. This is called **FLASHING**.

Reprogramming or "**Chipping**" offers instant transformations in how your vehicle performs; not only how the engine operates, but improved shift points as well.

Visit the resource page for a list of Chip Manufacturers.

www.HowToGasMileage.com/resources

Hydrogen Generators

Last, we have hydrogen generators. This is getting a little bit out of the scope of this book, and is a bit more involved, but does offer some benefits that I thought you may want to know about.

Some of the original alternative energy cars were based on **"Hydrogen Fuel Cells"**.

Basically, we're talking about water.

These hydrogen generators or "**water injection systems**" are available to purchase, and are even pretty easy to build. In essence, what they do is inject a small amount of water into your fuel intake system which changes the characteristics of the gas, creating more power.

More power means less gas needed.

Visit the resource page for a list of Hydrogen Products.

www.HowToGasMileage.com/resources

Choosing A Fuel Efficient Vehicle

Although the theme of this book is how to get the best results out of the vehicle you all ready own, what if you are considering getting a more economical vehicle?

I wanted to make sure you are armed with good information while making your decision.

Do yourself a favor and put some thought into how you intend to use this vehicle.

- Is this a second vehicle just for transportation?

- Are you replacing a family vehicle?

- Will you need to tow anything such as a camper?

The answers to these questions will have a impact on what type of vehicle you should consider purchasing.

Transmission

Standard transmissions or "stick shifts," get better gas mileage than automatic transmissions do. But keep in mind, drinking coffee while driving a stick shift can be a painful experience.

If you choose an automatic transmission, make sure it has OVERDRIVE.

Engine Size

For the most part, engine size has the biggest impact on a car's fuel economy.

A 4 cylinder engine will get the best gas mileage, providing it meets the needs of your intended use, such as a small passenger car.

Next of course is a 6 cylinder engine. These are often found in SUVs and Minivans, because they are heavier and often carry more passengers.

Are you planning on towing anything?

SUV 's can easily haul smaller trailers such as a pop-up camper, or a small boat. There are even light weight travel trailers made just for SUVs. In any case, you would want a 6 cylinder engine at a minimum for sure.

If you're planning on towing anything larger, such as a standard size travel trailer, a horse trailer, or an additional vehicle, you need an 8 cylinder engine.

As a general rule, you should not tow anything bigger than a jet ski with a front-wheel-drive vehicle! No matter what the salesman says, or even the owner's manual states.

I say this from being in the transmission repair business for over a decade. This is because front wheel drive transmissions are not built strong enough to handle the extra abuse of towing! They create too much heat, and that is the main cause of transmission failure.

Avoid the $3,000 transmission repair, get the right vehicle for the job.

Diesel engines get great gas mileage. However, they are often more expensive to purchase and repair. On top of that, diesel fuel usually costs more than gasoline.

Alternative Energy Vehicles

Hybrid vs. Electric

Alternative Energy Vehicles basically come in two versions, **Hybrid** and **Electric**. First, let's talk about electric vehicles.

Electric vehicles are just that, **electric**. They run on a battery just like a golf cart. Although the concept sounds nice, in reality, they are not always the most practical means of transportation.

First, they require electricity to charge them up. It's a known problem that our country's power grid is having a hard time keeping up with demand. Creating vehicles that need to be plugged in every night does not help this problem.

These vehicles also require large batteries. These batteries lose considerable power when they get cold. Also, these batteries are made out of very dangerous materials which are not environmentally friendly.

Now it may sound like I'm all against the electric vehicle, I'm not!

The electric motor is the most efficient power source we have. All I am saying is as a replacement for your normal transportation needs, it just isn't practical at this time.

However, if you can get over the large price tag, only need to drive it short distances, you live in the city, and it's a warm climate, then an electric vehicle may work good for you.

It would be better if you could use strictly solar power to keep it charged.

Now Hybrid Vehicles I do like!

These systems are based on the same technology that trains and ships have successfully used for years.

A Hybrid Vehicle has an electric motor and a very small gas engine. The gas engine's only job is to power a generator that provides electricity for the electric motor. The electric motor does the job of making the car go. So in other words, an electric motor does the driving, and a smaller gas motor generates the electricity it needs to do it.

Pretty Cool!

In my opinion, If you're looking into buying an Alternative Energy Vehicle,

A Hybrid is the only way to go.

Conclusion

The fact is, as individuals we have very little control over the price of gas. Sure we can all boycott and park our cars for month, but that's not likely to happen. The good news is that by utilizing the techniques I've laid out in this book, you **CAN** do something about your gas expenses.

As you go forward, make sure you take the time to track your expenses. Good tracking is the only way you will be able to see the fruits of your labor.

As you start seeing results, you will get excited, and become even more vigilant. Saving gas will become a competition and you'll want to beat your high score.

I hope you have enjoyed yourself while going through this book. I know I enjoyed writing it. I wish you all the best of luck on saving money at the pump.

Additional Resources

For additional information on the things covered in this publication,

visit

www.HowToGasMileage.com/resources